Book Cover & Illustrations by Danika Runyan

My Weather Book

TORNADO SAFETY

Written by **Matthew Jones**

Illustrated by **Danika Runyan**

To my little tiny tornado
and son, Grant.

Leo loved learning about the weather! His class had been studying storms all week. His teacher, Ms. Lily, told them, "The National Weather Service watches the skies every day and night. If they think storms might happen, they let us know by sending alerts and warnings."

"They look for weather conditions that could create dangerous storms, like tornadoes."

"What's a tornado?" Leo asked.
"A tornado," Ms. Lily explained, "is a spinning funnel of air that reaches down from a storm cloud to the ground."

Ms. Lily told the class,
"If a tornado spins your way,
don't be scared, be prepared!"

Leo was excited to tell his family all about storms!

When Leo got home, he said, "When a thunderstorm begins to spin, we call that storm a supercell. Tornadoes can form under spinning supercells."

One afternoon, the weather woman announced,
"A tornado watch has been issued for our county."

"A tornado watch means weather conditions are right for
tornadoes to form," Leo remembered.
"We need to be ready."

TORNADO WATCH

Leo and his family began to get their safe place ready.

It was a bathroom on the lowest level of their home,
away from outside walls and windows.

They gathered their emergency kit: a flashlight, a battery-powered radio,

water, snacks, blankets, and a first aid kit.

Leo grabbed his baseball helmet to protect his head and put on his shoes to protect his feet.

Leo remembered what Ms. Lily had said: "Don't be scared, be prepared!"

Later, the sky outside turned a greenish color.
The clouds looked dark and angry.
The storm was heading their way.

Suddenly, Leo's mom's phone buzzed loudly with a tornado alert. At the same time, they heard a loud wailing sound outside.

"Tornado sirens!" Dad said. "Those alerts mean the National Weather Service has issued a tornado warning."

"A tornado warning means a tornado has been spotted by a person or detected by weather radar," Mom said.
"We need to get to our safe place!"

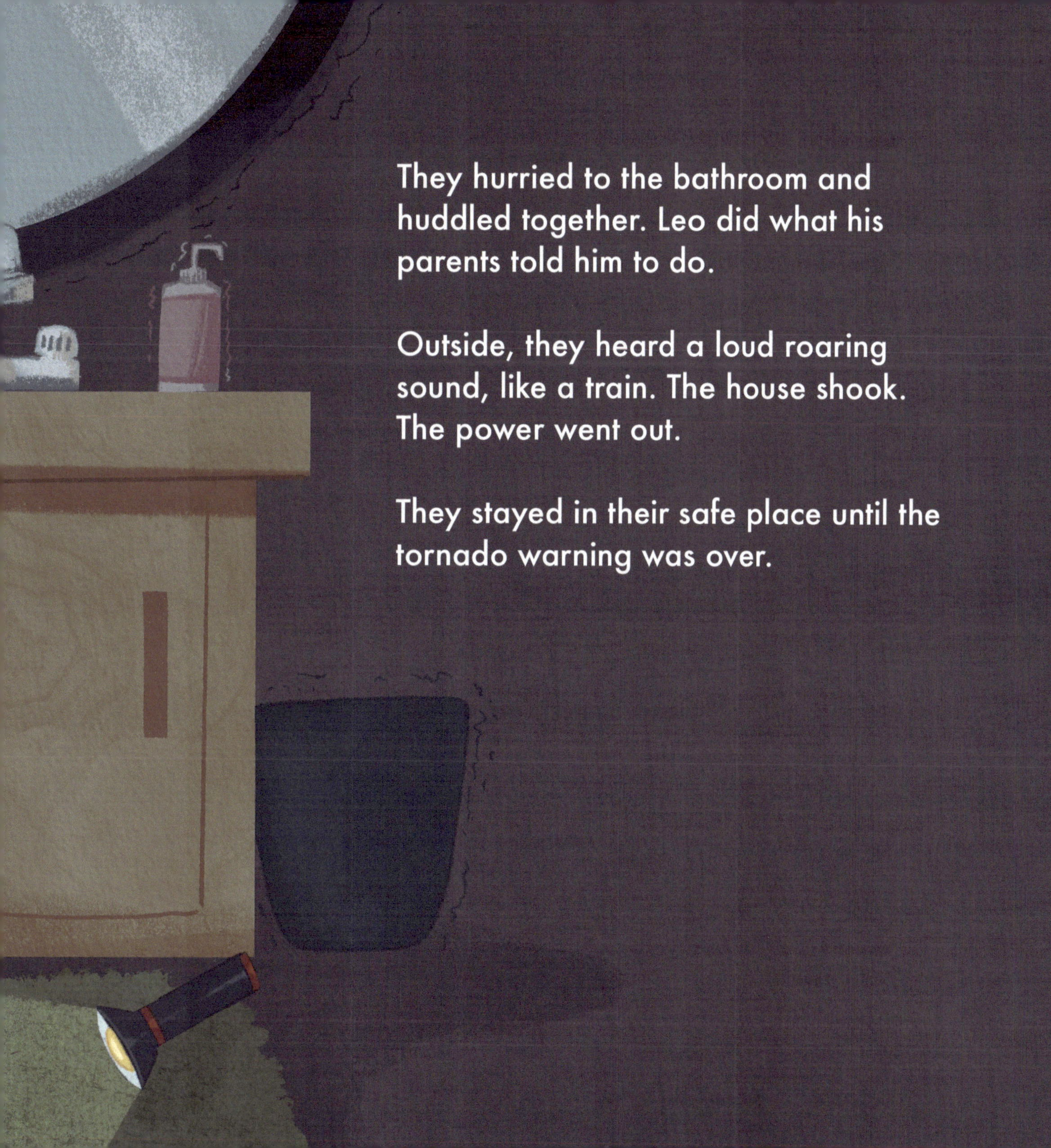

They hurried to the bathroom and huddled together. Leo did what his parents told him to do.

Outside, they heard a loud roaring sound, like a train. The house shook. The power went out.

They stayed in their safe place until the tornado warning was over.

POLICE

When they carefully went outside, Leo saw that a tornado had hit nearby. Trees had been knocked down and homes were damaged. He saw rescue crews—firefighters and police—helping people.

If your home is hit by a tornado, it's important to listen to grown-ups and wait for help. Don't touch anything broken.

Thankfully, their house was okay. Leo felt a little shaken,
but he remembered what Ms. Lily had said:
"Don't be scared, be prepared!"
Knowing what to do helps keep you safe.

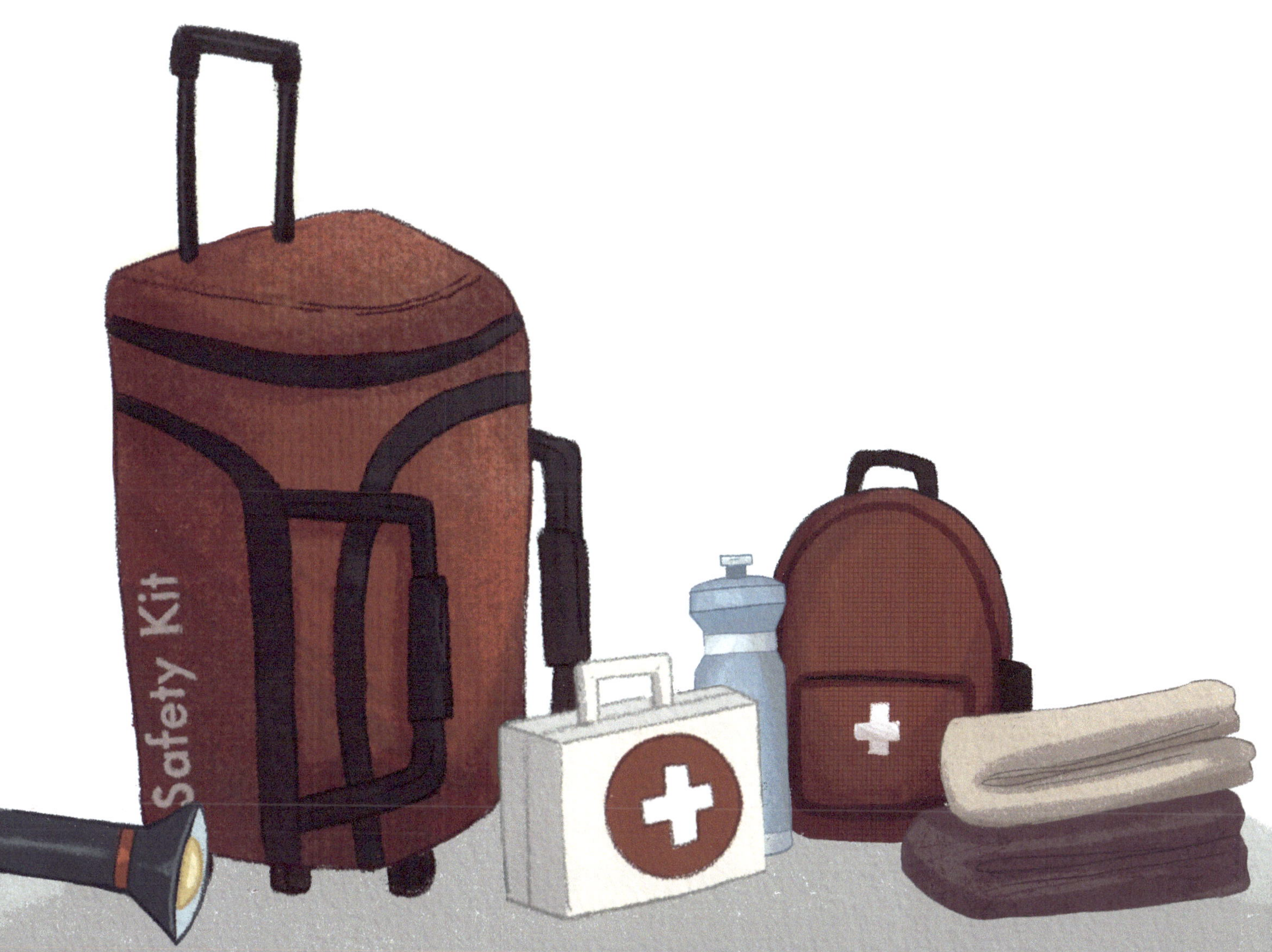

Leo was glad he knew what to do.
Being prepared had made a dangerous situation a little less scary.
He knew that even though big storms can happen,
being ready makes a big difference.

Let's talk more about tornado safety.
Here are some questions to ask your
friends and family:

1) What is the difference between a tornado watch and a tornado warning?

2) Do we know where we live on a map?

3) Do we have weather alerts on our phones turned on?

4) Where is our tornado safe place?

5) If we don't have a tornado safe place, do we know where the closest tornado shelter is?

6) What should we bring to our tornado safe place?

7) What do we do if there is a tornado warning at school?

8) Can you finish the sentence? Don't be scared, be _______!

Find the answers to these questions and more information about tornadoes at www.myweatherbook.com